I0075127

The Genetic Effects of Radiation

Contents

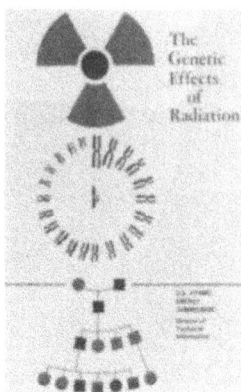

THE COVER

The cover design embodies a radiation symbol, a stylized karyotype of human chromosomes, and a genealogical table.

THE AUTHORS

ISAAC ASIMOV received his academic degrees from Columbia University and is Associate Professor of Biochemistry at the Boston University School of Medicine. He is a prolific author who has written over 83 books in the past 18 years, including about 20 science fiction works, and books for children. His many excellent science books for the public cover subjects in mathematics, physics, astronomy, chemistry, and biology, such as *The Genetic Code, Inside the Atom, Building Blocks of the Universe, The Living River, The New Intelligent Man's Guide to Science,* and *Asimov's Biographical Encyclopedia of Science and Technology.* In 1965 Dr. Asimov received the James T. Grady Award of the American Chemical Society for his major contribution in reporting science progress to the public.

THEODOSIUS DOBZHANSKY was graduated from Kiev University and is now a professor at the Rockefeller University. He has done research in genetics and biological evolution on every continent except Antarctica. Among his distinguished published works are *Radiation, Genes, and Man, Heredity and the Nature of Man, Mankind Evolving,* and *Evolution, Genetics, and Man.* Mr. Dobzhansky received the Daniel G. Elliot Prize and Medal and the Kimber Genetics Award from the National Academy of Sciences in 1958, and the National Medal of Science awarded by the President of the United States, in 1965.

The Genetic Effects of Radiation

By ISAAC ASIMOV
and
THEODOSIUS DOBZHANSKY

THE MACHINERY OF INHERITANCE

Introduction

There is nothing new under the sun, says the Bible. Nor is the sun itself new, we might add. As long as life has existed on earth, it has been exposed to radiation from the sun, so that life and radiation are old acquaintances and have learned to live together.

We are accustomed to looking upon sunlight as something good, useful, and desirable, and certainly we could not live long without it. The energy of sunlight warms the earth, produces the winds that tend to equalize earth's temperatures, evaporates the oceans and produces rain and fresh water. Most important of all, it supplies what is needed for green plants to convert carbon dioxide and water into food and oxygen, making it possible for all animal life (including ourselves) to live.

Yet sunlight has its dangers, too. Lizards avoid the direct rays of the noonday sun on the desert, and we ourselves take precautions against sunburn and sunstroke.

The same division into good and bad is to be found in connection with other forms of radiation—forms of which mankind has only recently become aware. Such radiations, produced by radioactivity in the soil and reaching us from outer space, have also been with us from the beginning of time. They are more energetic than sunlight, however, and can do more damage, and because our senses do not detect them, we have not learned to take precautions against them.

To be sure, energetic radiation is present in nature in only very small amounts and is not, therefore, much of a danger. Man, however, has the capacity of imitating nature. Long ago in dim prehistory, for instance, he learned to manufacture a kind of sunlight by setting wood and other fuels on fire. This involved a new kind of good and bad. A whole new technology became possible, on the one hand, and, on the other, the chance of death by burning was also possible. The good in this case far outweighs the evil.

In our own twentieth century, mankind learned to produce energetic radiation in concentrations far surpassing those we usually encounter in nature. Again, a new technology is resulting and again there is the possibility of death.

The balance in this second instance is less certainly in favor of the good over the evil. To shift the balance clearly in favor of the good, it is necessary for mankind to learn as much as possible about the new dangers in order that we might minimize them and most effectively guard against them.

To see the nature of the danger, let us begin by considering living tissue itself — the living tissue that must withstand the radiation and that can be damaged by it.

Cells and Chromosomes

The average human adult consists of about 50 trillion *cells* — 50 trillion microscopic, more or less self-contained, blobs of life. He begins life, however, as a single cell, the *fertilized ovum.*

After the fertilized ovum is formed, it divides and becomes two cells. Each daughter cell divides to produce a total of four cells, and each of those divides and so on.

There is a high degree of order and direction to those divisions. When a human fertilized ovum completes its divisions an adult human being is the inevitable result. The fertilized ovum of a giraffe will produce a giraffe, that of a fruit fly will produce a fruit fly, and so on. There

2

are no mistakes, so it is quite clear that the fertilized ovum must carry "instructions" that guide its development in the appropriate direction.

These "instructions" are contained in the cell's *chromosomes,* tiny structures that appear most clearly (like stubby bits of tangled spaghetti) when the cell is in the actual process of division. Each species has some characteristic number of chromosomes in its cells, and these chromosomes can be considered in pairs. Human cells, for instance, contain 23 pairs of chromosomes—46 in all.

When a cell is undergoing division (*mitosis*), the number of chromosomes is temporarily doubled, as each chromosome brings about the formation of a replica of itself. (This process is called *replication.*) As the cell divides, the chromosomes are evenly shared by the new cells in such a way that if a particular chromosome goes into one

Mitosis

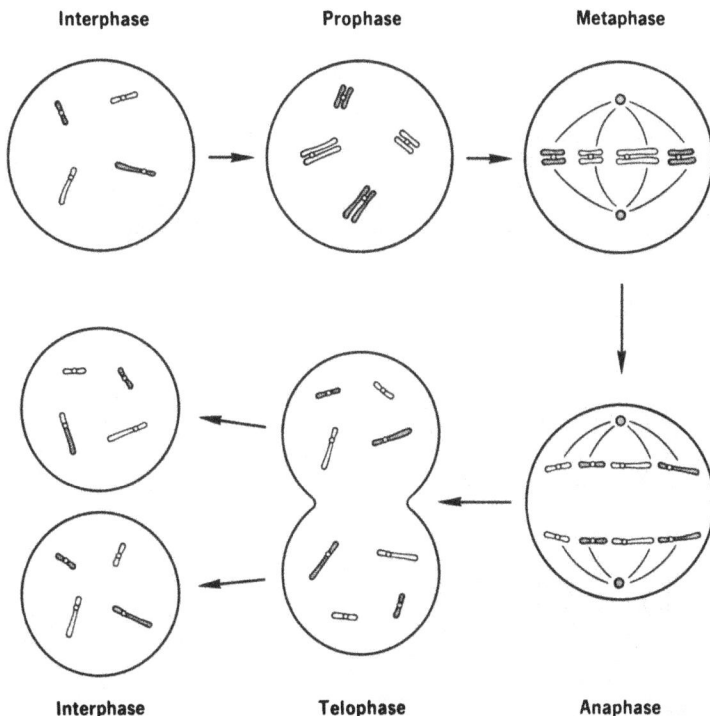

Interphase   Prophase   Metaphase

Interphase   Telophase   Anaphase

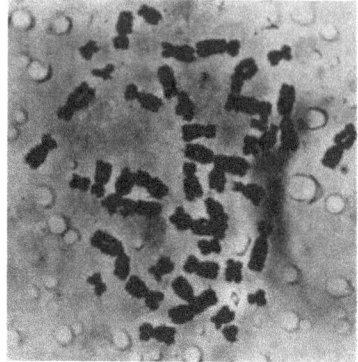

To study chromosomes, scientists begin with a cell that is in the process of dividing, when chromosomes are in their most visible form. Then they treat the cell with a chemical, a derivative of colchicine, to arrest the cell division at the metaphase stage (see mitosis diagram on preceding page). This brings a result like the photomicrograph at left above; the chromosomes are visible but still too tangled to be counted or measured. Then the cell is treated with a low-concentration salt solution, which swells the chromosomes and disperses them so they become distinct structures, as on the right.

The separate chromosomes in a dividing cell are photographed and then can be identified by their overall length, the position of the centromere, or point where the two strands join, and other characteristics. The photomicrograph can then be cut apart and the chromosomes grouped in a karyotype, which is an arrangement according to a standard classification to show chromosome complement and abnormalities. The karotype below is of a normal male, since it shows X and Y sex chromosomes and 22 pairs of other, autosomal, chromosomes. By contrast, the cells in the upper pictures are abnormal, with only 45 chromosomes each.

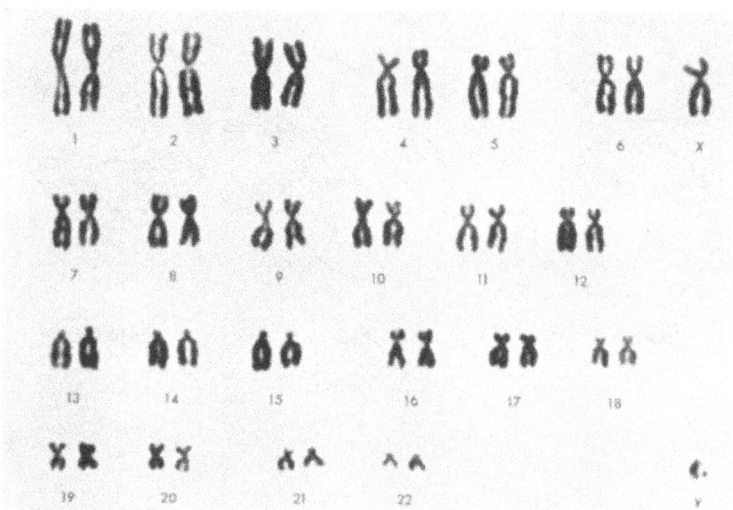

daughter cell, its replica goes into the other. In the end, each cell has a complete set of pairs of chromosomes; and the set in each cell is identical with the set in the original cell before division.

In this way, the fundamental "instructions" that determine the characteristics of a cell are passed on to each new cell. Ideally, all the trillions of cells in a particular human being have identical sets of "instructions".*

Enzymes and Genes

Each cell is a tiny chemical factory in which several thousand different kinds of chemical changes are constantly taking place among the numerous sorts of molecules that move about in its fluid or that are pinned to its solid structures. These chemical changes are guided and controlled by the existence of as many thousands of different *enzymes* within the cell.

Enzymes possess large molecules built up of some 20 different, but chemically related, units called *amino acids*. A particular enzyme molecule may contain a single amino acid of one type, five of another, several dozen of still another and so on. All the units are strung together in some specific pattern in one long chain, or in a small number of closely connected chains.

Every different pattern of amino acids forms a molecule with its own set of properties, and there are an enormous number of patterns possible. In an enzyme molecule made up of 500 amino acids, the number of possible patterns can be expressed by a 1 followed by 1100 zeroes (10^{1100}).

Every cell has the capacity of choosing among this unimaginable number of possible patterns and selecting those characteristic of itself. It therefore ends with a complement of specific enzymes that guide its own chemical changes and, consequently, its properties and its behavior. The "instructions" that enable a fertilized ovum to develop in the proper manner are essentially "instructions" for choosing a particular set of enzyme patterns out of all those possible.

---

*For more detail about cell division, see *Radioisotopes and Life Processes,* another booklet in this series.

The differences in the enzyme-guided behavior of the cells making up different species show themselves in differences in body structure. We cannot completely follow the long and intricate chain of cause-and-effect that leads from one set of enzymes to the long neck of a giraffe and from another set of enzymes to the large brain of a man, but we are sure that the chain is there. Even within a species, different individuals will have slight distinctions among their sets of enzymes and this accounts for the fact that no two human beings are exactly alike (leaving identical twins out of consideration).

Each chromosome can be considered as being composed of small sections called *genes,* usually pictured as being strung along the length of the chromosome. Each gene is considered to be responsible for the formation of a chain of amino acids in a fixed pattern. The formation is guided by the details of the gene's own structure (which are the "instructions" earlier referred to). This gene structure, which can be translated into an enzyme's structure, is now called the *genetic code.*

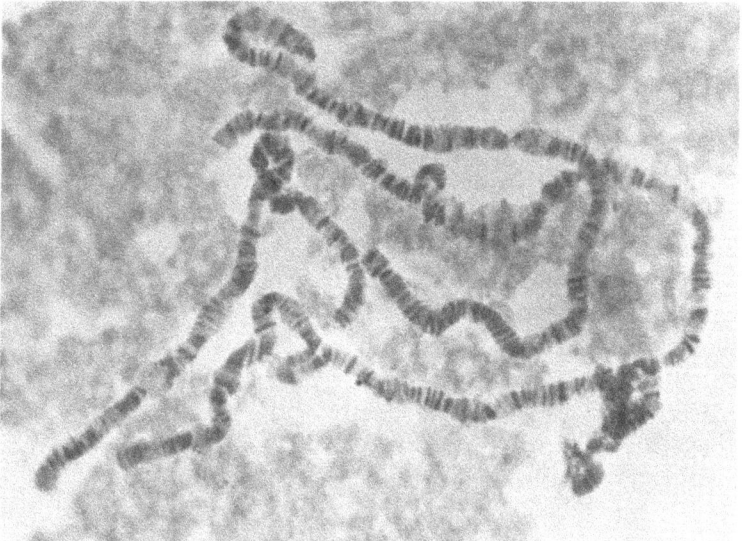

Stained section of one cell from salivary gland of Drosophila, *or fruit flies, reveals dark bands that may be genes controlling specific traits.*

If a particular enzyme (or group of enzymes) is, for any reason, formed imperfectly or not at all, this may show up as some visible abnormality of the body—an inability to see color, for instance, or the possession of two joints in each finger rather than three. It is much easier to observe physical differences than some delicate change in the enzyme pattern of the cells. Genes are therefore usually referred to by the body change they bring about, and one can, for instance, speak of a "gene for color blindness".

A gene may exist in two or more varieties, each producing a slightly different enzyme, a situation that is reflected, in turn, in slight changes in body characteristics. Thus, there are genes governing eye color, one of which is sufficiently important to be considered a "gene for blue eyes" and another a "gene for brown eyes". One or the other, but not both, will be found in a specific place on a specific chromosome.

The two chromosomes of a particular pair govern identical sets of characteristics. Both, for instance, will have a place for genes governing eye color. If we consider only the most important of the varieties involved, those on each chromosome of the pair may be identical; both may be for blue eyes or both may be for brown eyes. In that case, the individual is *homozygous* for that characteristic and may be referred to as a *homozygote*. The chromosomes of the pair may carry different varieties: A gene for blue eyes on one chromosome and one for brown eyes on the other. The individual is then *heterozygous* for that characteristic and may be referred to as a *heterozygote*. Naturally, particular individuals may be homozygous for some types of characteristics and heterozygous for others.

When an individual is heterozygous for a particular characteristic, it frequently happens that he shows the effect associated with only one of the gene varieties. If he possesses both a gene for brown eyes and one for blue eyes, his eyes are just as brown as though he had carried two genes for brown eyes. The gene for brown eyes is *dominant* in this case while the gene for blue eyes is *recessive*.

Parents and Offspring

How does the fertilized ovum obtain its particular set of chromosomes in the first place?

Each adult possesses gonads in which *sex cells* are formed. In the male, sperm cells are formed in the testes; in the female, egg cells are formed in the ovaries.

In the formation of the sperm cells and egg cells there is a key step — *meiosis* — a cell division in which the chromosomes group into pairs and are then apportioned between the daughter cells, one of each pair to each cell. Such a division, unaccompanied by replication, means that in place of the usual 23 pairs of chromosomes in each other cell, each sex cell has 23 individual chromosomes, a "half-set", so to speak.

In the process of fertilization, a sperm cell from the father enters and merges with an egg cell from the mother. The fertilized ovum that results now has a full set of 23 pairs of chromosomes, but of each pair, one comes from the father and one from the mother.

In this way, each newborn child is a true individual, with its characteristics based on a random reshuffling of chromosomes. In forming the sex cells, the chromosome pairs can separate in either fashion (*a* into cell 1 and *b* into cell 2, or vice versa). If each of 23 pairs does this randomly, nearly 10 million different combinations of chromosomes are possible in the sex cells of a single individual.

Furthermore, one can't predict which chromosome combination in the sperm cell will end up in combination with which in the egg cell, so that by this reasoning, a single married couple could produce children with any of 100 trillion (100,000,000,000,000) possible chromosome combinations.

It is this that begins to explain the endless variety among living beings, even within a particular species.

It only begins to explain it, because there are other sources of difference, too. A chromosome is capable of exchanging pieces with its pair, producing chromosomes with a brand new pattern of gene varieties. Before such a *crossover*, one chromosome may have carried a gene for

Meiosis

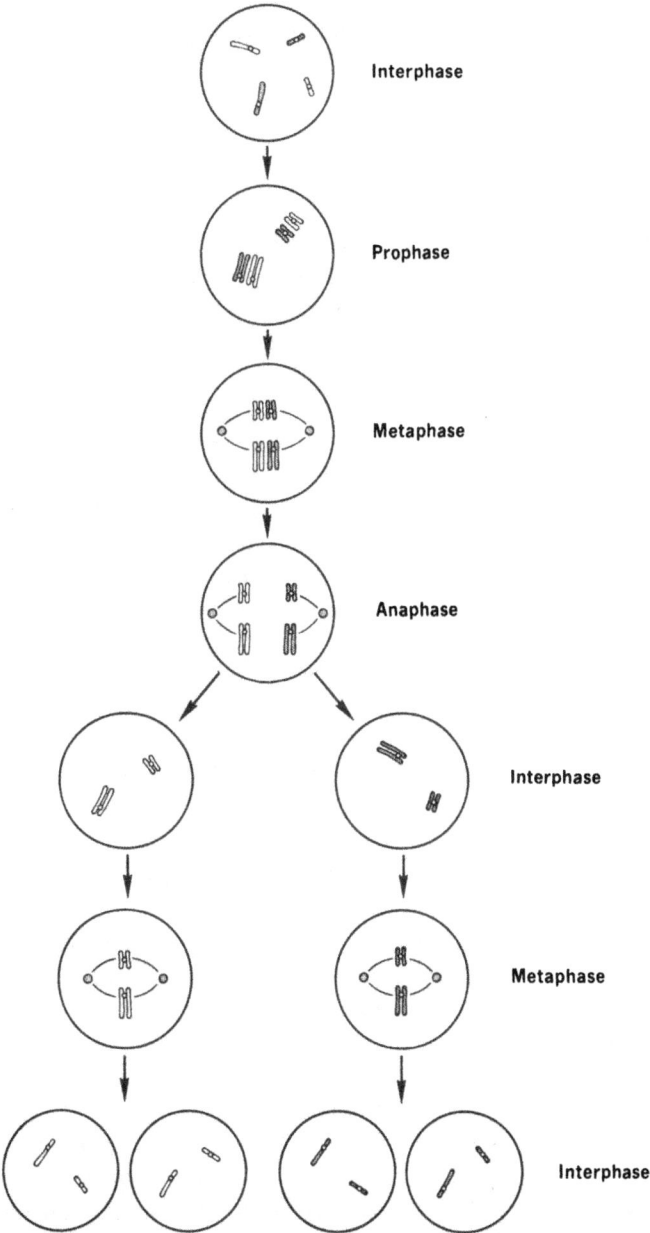

Interphase

Prophase

Metaphase

Anaphase

Interphase

Metaphase

Interphase

9

blue eyes and one for wavy hair, while the other chromosome may have carried a gene for brown eyes and one for straight hair. After the crossover, one would carry genes for blue eyes and straight hair, the other for brown eyes and wavy hair.

MUTATIONS

Sudden Change

Shifts in chromosome combinations, with or without crossovers, can produce unique organisms with characteristics not quite like any organism that appeared in the past nor likely to appear in the reasonable future. They may even produce novelties in individual characteristics since genes can affect one another, and a gene surrounded by unusual neighbors can produce unexpected effects.

Matters can go further still, however, in the direction of novelty. It is possible for chromosomes to undergo more serious changes, either structural or chemical, so that entirely new characteristics are produced that might not otherwise exist. Such changes are called *mutations*.

We must be careful how we use this term. A child may possess some characteristics not present in either parent through the mere shuffling of chromosomes and not through mutation.

Suppose, for instance, that a man is heterozygous to eye color, carrying one gene for brown eyes and one for blue eyes. His eyes would, of course, be brown since the gene for brown eyes is dominant over that for blue. Half the sperm cells he produces would carry a single gene for brown eyes in its half set of chromosomes. The other half would carry a single gene for blue eyes. If his wife were similarly heterozygous (and therefore also had brown eyes), half her egg cells would carry the gene for brown eyes and half the gene for blue.

It might follow in this marriage, then, that a sperm carrying the gene for blue eyes might fertilize an egg carrying the gene for blue eyes. The child would then be homozygous, with two genes for blue eyes, and he would definitely be blue-eyed. In this way, two brown-eyed parents might have a blue-eyed child and this would *not* be a mutation. If the parents' ancestry were traced further back, blue-eyed individuals would undoubtedly be found on both sides of the family tree.

If, however, there were no record of, say, anything but normal color vision in a child's ancestry, and he were born color-blind, that could be assumed to be the result of a mutation. Such a mutation could then be passed on by the normal modes of inheritance and a certain proportion of the child's eventual descendants would be color-blind.

A mutation may be associated with changes in chromosome structure sufficiently drastic to be visible under the microscope. Such *chromosome mutations* can arise in several ways. Chromosomes may undergo replication without the cell itself dividing. In that way, cells can develop with two, three, or four times the normal complement of chromosomes, and organisms made up of cells displaying such *polyploidy* can be markedly different from the norm. This situation is found chiefly among plants and among some groups of invertebrates. It does not usually occur in mammals, and when it does it leads to quick death.

Less extreme changes take place, too, as when a particular chromosome breaks and fails to reunite, or when several break and then reunite incorrectly. Under such conditions, the mechanism by which chromosomes are distributed among the daughter cells is not likely to work correctly. Sex cells may then be produced with a piece of chromosome (or a whole one) missing, or with an extra piece (or whole chromosome) present.

In 1959, such a situation was found to exist in the case of persons suffering from a long-known disease called Down's syndrome.* Each person so afflicted has 47 chromosomes in place of the normal 46. It turned out that the 21st pair of chromosomes (using a convention whereby the chromosome pairs are numbered in order of decreasing size) consists of three individuals rather than two. The existence of this chromosome abnormality clearly demonstrated what had previously been strongly suspected—that Down's syndrome originates as a mutation and is inborn (see the figure on the next page).

---

*This is more commonly known as "Mongolism" or "Mongolian idiocy" though it has nothing to do with the Mongolian people.

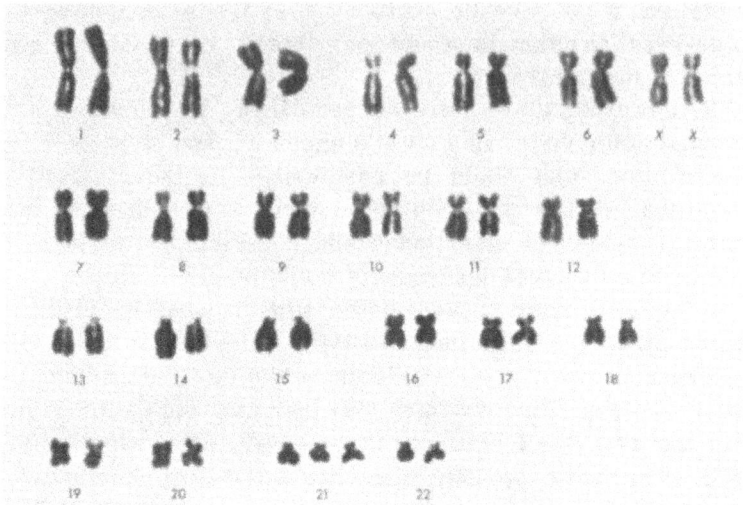

Karyotype of a female patient with Down's syndrome (Mongolism). During meiosis both chromosomes No. 21 of the mother, instead of just one, went to the ovum. Fertilization added the father's chromosome, which made three Nos. 21 instead of the normal pair. (Compare with the normal karyotype on page 4.)

Most mutations, however, are not associated with any noticeable change in chromosome structure. There are, instead, more subtle changes in the chemical structure of the genes that make up the chromosome. Then we have *gene mutations.*

The process by which a gene produces its own replica is complicated and, while it rarely goes wrong, it does misfire on occasion. Then, too, even when a gene molecule is replicated perfectly, it may undergo change afterward through the action upon it of some chemical or other environmental influence. In either case, a new variety of a particular gene is produced and, if present in a sex cell, it may be passed on to descendants through an indefinite number of generations.

Of course, chromosome or gene mutations may take place in ordinary cells rather than in sex cells. Such changes in ordinary cells are *somatic mutations.* When mutated body cells divide, new cells with changed characteristics are produced. These changes may be trivial, or they may be serious. It is often suggested, for instance, that cancer may result from a somatic mutation in which certain cells lose the capacity to regulate their growth properly. Since somatic mutations do not involve the sex cells, they are confined to the individual and are not passed on to the offspring.

Spontaneous Mutations

Mutations that take place in the ordinary course of nature, without man's interference, are *spontaneous mutations*. Most of these arise out of the very nature of the complicated mechanism of gene replication. Copies of genes are formed out of a large number of small units that must be lined up in just the right pattern to form one particular gene and no other.

Ideally, matters are so arranged within the cell that the necessary changes giving rise to the desired pattern are just those that have a maximum probability. Other changes are less likely to happen but are not absolutely excluded. Sometimes through the accidental jostling of molecules a wrong turn may be taken, and the result is a spontaneous mutation.

We might consider a mutation to be either "good" or "bad" in the sense that any change that helps a creature live more easily and comfortably is good and that the reverse is bad.

It seems reasonable that random changes in the gene pattern are almost sure to be bad. Consider that any creature, including man, is the product of millions of years of evolution. In every generation those individuals with a gene pattern that fit them better for their environment won out over those with less effective patterns — won out in the race for food, for mates, and for safety. The "more fit" had more offspring and crowded out the "less fit".

By now, then, the set of genes with which we are normally equipped is the end product of long ages of such *natural selection*. A random change cannot be expected to improve it any more than random changes would improve any very complex, intricate, and delicate structure.

Yet over the eons, creatures have indeed changed, largely through the effects of mutation. If mutations are

Pleistocene and Recent

Pliocene

Miocene

Oligocene

Eocene

Evolution of the horse (skull, hindfoot, and forefoot shown). Note the changes over a 60-million-year period from the Eocene era to the present.

almost always for the worse, how can one explain that evolution seems to progress toward the better and that out of a primitive form as simple as an amoeba, for instance, there eventually emerged man.

In the first place, environment is not fixed. Climate changes, conditions change, the food supply may change, the nature of living enemies may change. A gene pattern that is very useful under one set of conditions may be less useful under another.

Suppose, for instance, that man had lived in tropical areas for thousands of years and had developed a heavily pigmented skin as a protection against sunburn. Any child who, through a mutation, found himself incapable of forming much pigment, would be at a severe disadvantage in the outdoor activities engaged in by his tribe. He would not do well and such a mutated gene would never establish itself for long.

If a number of these men migrated to northern Europe, however, children with dark skin would absorb insufficient sunlight during the long winter when the sun was low in the sky, and visible for brief periods only. Dark-skinned children would, under such conditions, tend to suffer from rickets.

Mutant children with pale skin would absorb more of what weak sunlight there was and would suffer less. There would be little danger of sunburn so there would be no penalty counteracting this new advantage of pale skins. It would be the dark-skinned people who would tend to die out. In the end, you would have dark skins in Africa and pale skins in Scandinavia, and both would be "fit".

In the same way, any child born into a primitive hunting society who found himself with a mutated gene that brought about nearsightedness would be at a distinct disadvantage. In a modern technological society, however, nearsighted individuals, doing more poorly at outdoor games, are often driven into quieter activities that involve reading, thinking, and studying. This may lead to a career as a scientist, scholar, or professional man, categories that are valuable in such a society and are encouraged. Nearsightedness would therefore spread more generally through civilized societies than through primitive ones.

Then, too, a gene may be advantageous when it occurs in low numbers and disadvantageous when it occurs in high numbers. Suppose there were a gene among humans that so affected the personality as to make it difficult for a human being to endure crowded conditions. Such individuals would make good explorers, farmers, and herdsmen, but poor city dwellers. Even in our modern urbanized society, such a gene in moderate concentration would be good, since we still need our outdoorsmen. In high concentration, it would be bad, for then the existence of areas of high population density (on which our society now seems to depend) might become impossible.

In any species, then, each gene exists in a number of varieties upon which an absolute "good" or "bad" cannot be unequivocally stamped. These varieties make up the *gene pool*, and it is this gene pool that makes evolution possible.

A species with an invariable set of genes could not change to suit altered conditions. Even a slight shift in the nature of the environment might suffice to wipe it out.

The possession of a gene pool lends flexibility, however. As conditions change, one combination of varieties might gain over another and this, in turn, might produce changes in body characteristics that would then further alter the relative "goodness" or "badness" of certain gene patterns.

Thus, over the past million years, for example, the human brain has, through mutations and appropriate shifts in emphasis within the gene pool, increased notably in size.

Genetic Load

Some gene mutations produce characteristics so undesirable that it is difficult to imagine any reasonable change in environmental conditions that would make them beneficial. There are mutations that lead to the nondevelopment of hands and feet, to the production of blood that will not clot, to serious malformations of essential organs, and so on. Such mutations are unqualifiedly bad.

The badness may be so severe that a fertilized ovum may be incapable of development; or, if it develops, the fetus miscarries or the child is stillborn; or, if the child is born alive, it dies before it matures so that it can never have children of its own. Any mutation that brings about death before the gene producing it can be passed on to another generation is a *lethal mutation.*

A gene governing a lethal characteristic may be dominant. It will then kill even though the corresponding gene on the other chromosome of the pair is normal. Under such conditions, the lethal gene is removed in the same generation in which it is formed.

The lethal gene may, on the other hand, be recessive. Its effect is then not evident if the gene it is paired with is normal. The normal gene carries on for both.

When this is the case, the lethal gene will remain in existence and will, every once in a while, make itself evident. If two people, each serving as a *carrier* for such a gene, have children, a sperm cell carrying a lethal may fertilize an egg cell carrying the same type of lethal, with sad results.

Every species, including man, includes individuals who carry undesirable genes. These undesirable genes may be passed along for generations, even if dominant, before natural selection culls them out. The more seriously undesirable they are, the more quickly they are removed, but even outright lethal genes will be included among the chromosomes from generation to generation provided they are recessive. These deleterious genes make up the *genetic load.*

The only way to avoid a genetic load is to have no mutations and therefore no gene pool. The gene pool is necessary for the flexibility that will allow a species to survive and evolve over the eons and the genetic load is the price that must be paid for that. Generally, the capacity for a species to reproduce itself is sufficiently high to make up, quite easily, the numbers lost through the combination of deleterious genes.

The size of a genetic load depends on two factors: The rate at which a deleterious gene is produced through mutation, and the rate at which it is removed by natural selection. When the rate of removal equals the rate of production, a condition of *genetic equilibrium* is reached and the level of occurrence of that gene then remains stable over the generations.

Even though deleterious genes are removed relatively rapidly, if dominant, and lethal genes are removed in the same generation in which they are formed, a new crop of deleterious genes will appear by mutation with every succeeding generation. The equilibrium level for such dominant deleterious genes is relatively low, however.

Deleterious genes that are recessive are removed much more slowly. Those persons with two such genes, who alone show the bad effects, are like the visible portion of an iceberg and represent only a small part of the whole. The heterozygotes, or carriers, who possess a single gene of this sort, and who live out normal lives, keep that gene in being. If people in a particular population marry randomly and if one out of a million is born homozygous for a certain deleterious recessive gene (and dies of it), one out of five hundred is heterozygous for that same gene, shows no ill effects, and is capable of passing it on.

It may be that the heterozygote is not quite normal but does show some ill effects—not enough to incommode him seriously, perhaps, but enough to lower his chances slightly for mating and bearing children. In that case, the equilibrium level for that gene will be lower than it would otherwise be.

It may also be that the heterozygote experiences an actual advantage over the normal individual under some conditions. There is a recessive gene, for instance, that produces a serious disease called sickle-cell anemia. People possessing two such genes usually die young. A heterozygote possessing only one of these genes is not seriously affected and has red blood cells that are, apparently, less appetizing to malaria parasites. The heterozygote therefore experiences a positive advantage if he lives in a region where the incidence of certain kinds of malaria is high. The equilibrium level of the sickle-cell anemia gene can, in other words, be higher in malarial regions than elsewhere.

Here is one subject area in which additional research is urgently needed. It may be that the usefulness of a single deleterious gene is greater than we may suspect in many cases, and that there are greater advantages to heterozygousness than we know. This may be the basis of what is sometimes called "hybrid vigor". In a world in which human beings are more mobile than they have ever been in history and in which intercultural marriages are increasingly common, information on this point is particularly important.

Mutation Rates

It is easier to observe the removal of genes through death or through failure to reproduce than to observe their production through mutation. It is particularly difficult to study their production in human beings, since men have comparatively long lifetimes and few children, and since their mating habits cannot well be controlled.

For this reason, geneticists have experimented with species much simpler than man — smaller organisms that are short-lived, produce many offspring, and that can be penned up and allowed to mate only under fixed conditions. Such creatures may have fewer chromosomes than man does and the sites of mutation are more easily pinned down.

An important assumption made in such experiments is that the machinery of inheritance and mutation is essentially the same in all creatures and that therefore knowledge gained from very simple species (even from bacteria) is applicable to man. There is overwhelming evidence to indicate that this is true in general, although there are specific instances where it is not completely true and scientists must tread softly while drawing conclusions.

The animals most commonly used in studies of genetics and mutations are certain species of fruit flies, called *Drosophila*. The American geneticist, Hermann J. Muller, devised techniques whereby he could study the occurrence of lethal mutations anywhere along one of the four pairs of chromosomes possessed by *Drosophila*.

A lethal gene, he found, might well be produced somewhere along the length of a particular chromosome once out of every two hundred times that chromosome underwent replication. This means that out of every 200 sex cells produced by *Drosophila,* one would contain a lethal gene somewhere along the length of that chromosome.

Geneticist Hermann J. Muller studying Drosophila *in his laboratory. Dr. Muller won a Nobel Prize in 1946 for showing that radiation can cause mutations. (See page 34.)*

That particular chromosome, however, contained at least 500 genes capable of undergoing a lethal mutation. If each of those genes is equally likely to undergo such a mutation, then the chance that any one particular gene is lethal is one out of 200×500, or 1 out of 100,000.

This is a typical mutation rate for a gene in higher organisms generally, as far as geneticists can tell (though the rates are lower among bacteria and viruses). Naturally, a chance for mutation takes place every time a new individual is born. Fruit flies have many more offspring per year than human beings, since their generations are shorter and they produce more young at a time. For that reason, though the mutation rate may be the same in fruit flies as in man, many more actual mutations are produced per unit time in fruit flies than in men.

This does not mean that the situation may be ignored in the case of man. Suppose the rate for production of a particular deleterious gene in man is 1 out of 100,000. It is estimated that a human being has at least 10,000 different genes, and therefore the chance that at least one of the genes in a sex cell is deleterious is 10,000 out of 100,000 or 1 out of 10.

Furthermore, it is estimated that the number of gene mutations that are weakly deleterious are four times as numerous as those that are strongly deleterious or lethal. The chances that at least one gene in a sex cell is at least weakly deleterious then would be 4 + 1 out of 10, or 1 out of 2.

Naturally, these deleterious genes are not necessarily spread out evenly among human beings with one to a sex cell. Some sex cells will be carrying more than one, thus increasing the number that may be expected to carry none at all. Even so, it is supposed that very nearly half the sex cells produced by humanity carry at least one deleterious gene.

Even though only half the sex cells are free of deleterious genes, it is still possible to produce a satisfactory new generation of men. Yet one can see that the genetic load is quite heavy and that anything that would tend to increase it would certainly be undesirable, and perhaps even dangerous.

We tend to increase the genetic load by reducing the rate at which deleterious genes are removed, that is, by taking care of the sick and retarded, and by trying to prevent discomfort and death at all levels.

There is, however, no humane alternative to this. What's more, it is, by and large, only those with slightly deleterious genes who are preserved genetically. It is those persons with nearsightedness, with diabetes, and so on, who, with the aid of glasses, insulin, or other props, can go on to live normal lives and have children in the usual numbers. Those with strongly deleterious genes either die despite all that can be done for them even today or, at the least, do not have a chance to have many children.

The danger of an increase in the genetic load rests more heavily, then, at the other end — at measures that (usually inadvertently or unintentionally) increase the rate of production of mutant genes. It is to this matter we will now turn.

RADIATION

Ionizing Radiation

Our modern technological civilization exposes mankind to two general types of genetic dangers unknown earlier: Synthetic chemicals (or unprecedentedly high concentrations of natural ones) absent in earlier eras, and intensities of energetic radiation equally unknown or unprecedented.

Chemicals can interfere with the process of replication by offering alternate pathways with which the cellular machinery is not prepared to cope. In general, however, it is only those cells in direct contact with the chemicals that are so affected, such as the skin, the intestinal linings, the lungs, and the liver (which is active in altering and getting rid of foreign chemicals). These may undergo somatic mutations, and an increased incidence of cancer in those tissues is among the drastic results of exposure to certain chemicals.

Such chemicals are not, however, likely to come in contact with the gonads where the sex cells are produced. While individual persons may be threatened by the manner in which the environment is being permeated with novel chemicals, the next generation is not affected in advance.

Radiation is another matter. In its broadest sense, radiation is any phenomenon spreading out from some source in all directions. Physically, such radiation may consist of waves or of particles.* Of the wave forms the two best-known are sound and electromagnetic radiations.

Sound carries very low concentrations of energy. This energy is absorbed by living tissue and converted into heat. Heat in itself can increase the mutation rate but the effect is a small one. The body has effective machinery for keeping its temperature constant and the gonads are not likely to suffer unduly from exposure to heat.

---

*Actually, all waves have some of the characteristics of particles and all particles have some of the characteristics of waves. Usually, however, the radiation is predominantly one or the other and little confusion arises under ordinary circumstances in speaking of waves and particles as though they were separate phenomena.

Electromagnetic radiation comes in a wide range of energies, with visible light (the best-known example of such radiation because we can detect it directly and with great sensitivity) about in the middle of the range. Electromagnetic radiations less energetic than light (such as infrared waves and microwaves) are converted into heat when absorbed by living tissue. The heat thus formed is sufficient to cause atoms and molecules to vibrate more rapidly, but this added vibration is not usually sufficient to pull molecules apart and therefore does not bring about chemical changes.

Light will bring about some chemical changes. It is energetic enough to cause a mixture of hydrogen and chlorine to explode. It will break up silver compounds and produce tiny black grains of metallic silver (the chemical basis of photography). Living tissue, however, is largely unaffected — the retina of the eye being one obvious exception.

Ultraviolet light, which is more energetic than visible light, correspondingly can bring about chemical changes more easily. It will redden the skin, stimulate the production of pigment, and break up certain steroid molecules to form vitamin D. It will even interfere with replication to some extent. At least there is evidence that persistent exposure to sunlight brings about a heightened tendency to skin cancer. Ultraviolet light is not very penetrating, however, and its effects are confined to the skin.

Electromagnetic radiations more energetic than ultraviolet light, such as X rays and gamma rays, carry sufficient concentrations of energy to bring about changes not only in molecules but in the very structure of the atoms making up those molecules.

Atoms consist of particles (electrons), each carrying a negative electric charge and circling a tiny centrally located nucleus, which carries a positive electric charge.

Ordinarily, the negative charges of the electrons just balance the positive charge on the nucleus so that atoms and molecules tend to be electrically neutral. An X ray or gamma ray, crashing into an atom, will, however, jar electrons loose. What is left of the atom will carry a positive electric charge with the charge size proportional to the number of electrons lost.

An atom fragment carrying an electric charge is called an *ion*. X rays and gamma rays are therefore examples of *ionizing radiation.*

Radiations may consist of flying particles, too, and if these carry sufficient energy they are also ionizing in character. Examples are *cosmic rays, alpha rays,* and *beta rays.* Cosmic rays are streams of positively charged nuclei, predominantly those of the element hydrogen. Alpha rays are streams of positively charged helium nuclei. Beta rays are streams of negatively charged electrons. The individual particles contained in these rays may be referred to as *cosmic particles, alpha particles,* and *beta particles,* respectively.

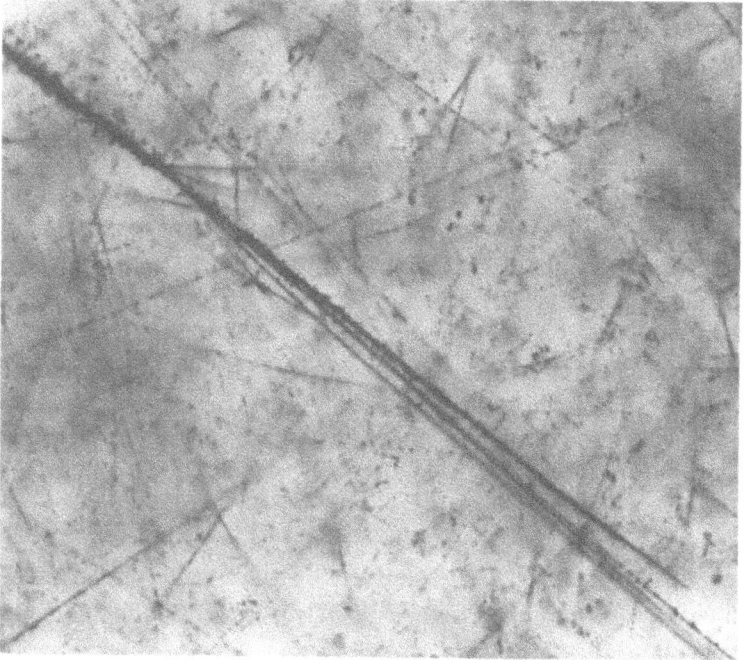

Cosmic ray and trapped Van Allen Belt energetic particles produced the dark tracks in this photo of a nuclear emulsion that had been carried aloft on an Air Force satellite. The energetic particles cause ionization of the silver bromide molecules in the emulsion.

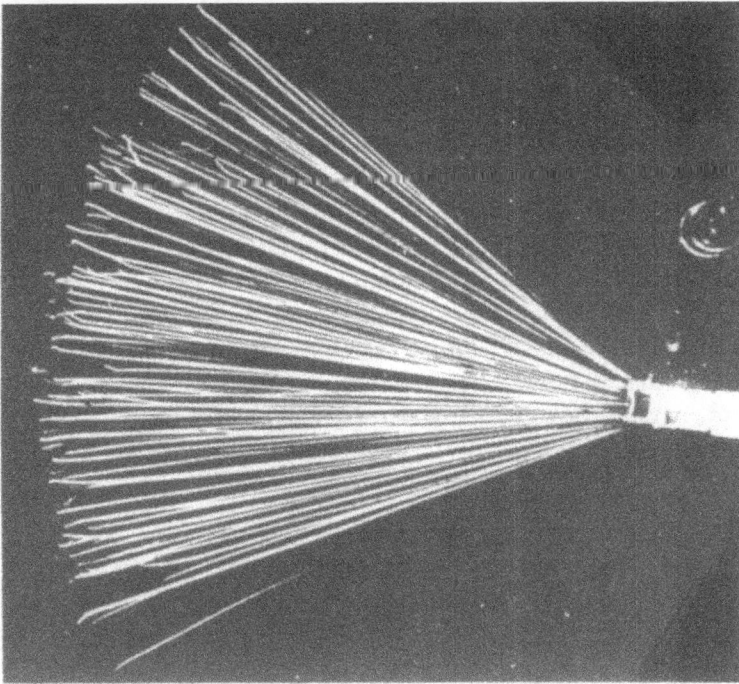

Alpha particles emitted by the source at right leave tracks in a cloud chamber. Some tracks are bent near the end as a result of collisions with atomic nuclei. Such collisions are more likely at the end of a track when the alpha particle has been slowed down.

Beta particles originating at left leave these tracks in a cloud chamber. Note that the tracks are much farther apart than those of alpha particles. As the particle slows down, its path becomes more erratic and the ions are formed closer together. At the very end of an electron track the proximity of the ions approximates that in an alpha-particle track.

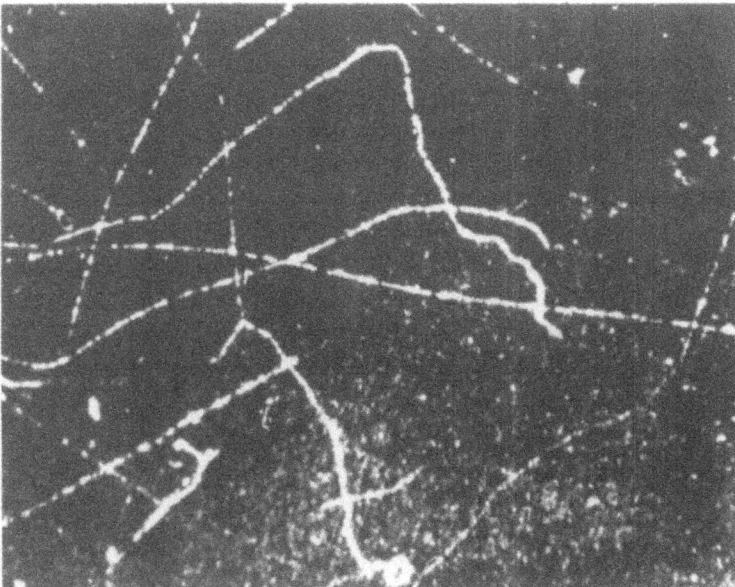

Ionizing radiation is capable of imparting so much energy to molecules as to cause them to vibrate themselves apart, producing not only ions but also high-energy uncharged molecular fragments called *free radicals*.

The direct effect of ionizing radiation on chromosomes can be serious. Enough chemical bonds may be disrupted so that a chromosome struck by a high-energy wave or particle may break into fragments. Even if the chromosome manages to remain intact, an individual gene along its length may be badly damaged and a mutation may be produced.

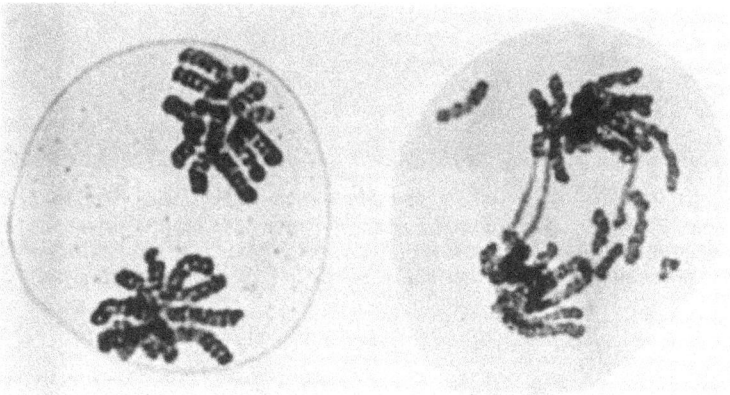

Effects of ionizing radiation on chromosomes: Left, a normal plant cell showing chromosomes divided into two groups; right, the same type of cell after X-ray exposure, showing broken fragments and bridges between groups, typical abnormalities induced by radiation.

If only direct hits mattered, radiation effects would be less dangerous than they are, since such direct hits are comparatively few. However, near-misses may also be deadly. A streaking bit of radiation may strike a water molecule near a gene and may break up the molecule to form a free radical. The free radical will be sufficiently energetic to bring about a chemical reaction with almost any molecule it strikes. If it happens to strike the neighboring gene before it has disposed of that energy, it will produce the mutation as surely as the original radiation might have.

Furthermore, ionizing radiations (particularly of the electromagnetic variety) tend to be penetrating, so that the interior of the body is as exposed as is the surface. The gonads cannot hide from X rays, gamma rays, or cosmic particles.

All these radiations can bring about somatic mutations — all can cause cancer, for instance.

What is worse, all of them increase the rate of genetic mutations so that their presence threatens generations unborn as well as the individuals actually exposed.

Background Radiation

Ionizing radiation in low intensities is part of our natural environment. Such natural radiation is referred to as *background radiation.* Part of it arises from certain constituents of the soil. Atoms of the heavy metals, uranium and thorium, are constantly, though very slowly, breaking down and in the process giving off alpha rays, beta rays, and gamma rays. These elements, while not among the most common, are very widely spread; minerals containing small quantities of uranium and thorium are to be found nearly everywhere.

In addition, all the earth is bombarded with cosmic rays from outer space and with streams of high-energy particles from the sun.

Various units can be used to measure the intensity of this background radiation. The *roentgen,* abbreviated *r*, and named in honor of the discoverer of X rays, Wilhelm Roentgen, is a unit based on the number of ions produced by radiation. Rather more convenient is another unit that has come more recently into prominence. This is the *rad* (an abbreviation for "radiation absorbed dose") that is a measure of the amount of energy delivered to the body upon the absorption of a particular dose of ionizing radiation. One rad is very nearly equal to one roentgen.

Since background radiation is undoubtedly one of the factors in producing spontaneous mutations, it is of interest to try to determine how much radiation a man or woman will have absorbed from the time he is first conceived to the time he conceives his own children. The average length

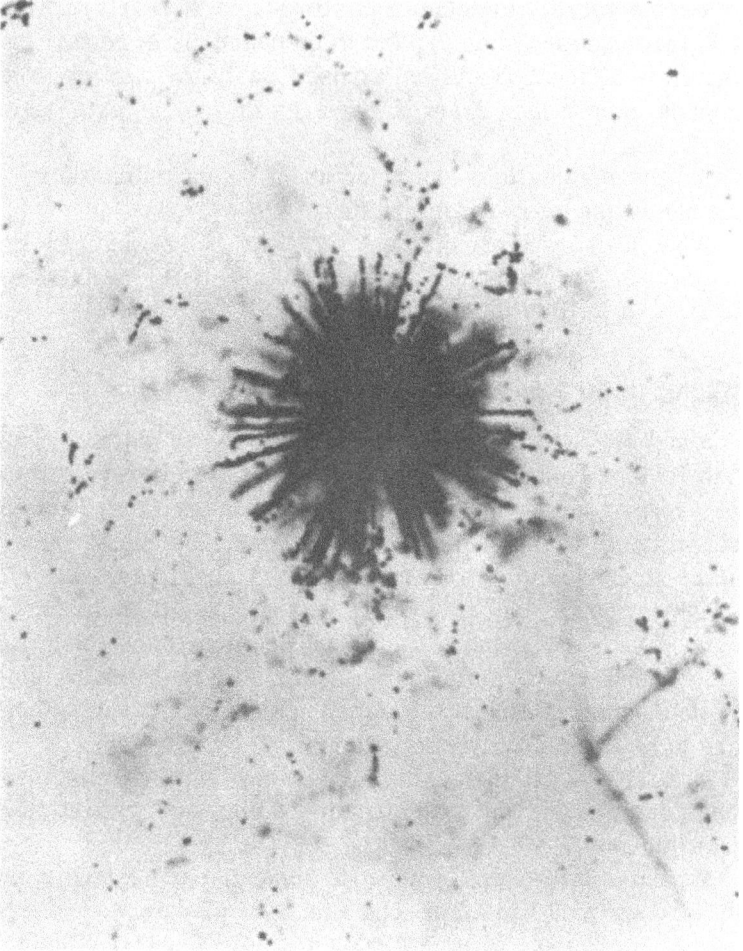

Natural radioactivity in the atmosphere is shown by this nuclear-emulsion photograph of alpha-particle tracks (enlarged 2000 diameters) emitted by a grain of radioactive dust.

of time between generations is taken to be about 30 years, so we can best express absorption of background radiation in units of *rads per 30 years*.

The intensity of background radiation varies from place to place on the earth for several reasons. Cosmic rays are deflected somewhat toward the magnetic poles by the earth's magnetic field. They are also absorbed by the atmosphere to some extent. For this reason, people living in equatorial regions are less exposed to cosmic rays than those in polar regions; and those in the plains, with a greater thickness of atmosphere above them, are less exposed than those on high plateaus.

Then, too, radioactive minerals may be spread widely, but they are not spread evenly. Where they are concentrated to a greater extent than usual, background radiation is abnormally high.

Thus, an inhabitant of Harrisburg, Pennsylvania, may absorb 2.64 rads per 30 years, while one of Denver, Colorado, a mile high at the foot of the Rockies, may absorb 5.04 rads per 30 years. Greater extremes are encountered at such places as Kerala, India, where nearby soil, rich in thorium minerals, so increases the intensity of background radiation that as much as 84 rads may be absorbed in 30 years.

In addition to high-energy radiation from the outside, there are sources within the body itself. Some of the potassium and carbon atoms of our body are inevitably radioactive. As much as 0.5 rad per 30 years arises from this source.

Rads and roentgens are not completely satisfactory units in estimating the biological effects of radiation. Some types of radiation—those made up of comparatively large particles, for instance—are more effective in producing ions and bring about molecular changes with greater ease than do electromagnetic radiations delivering equal energy to the body. Thus if 1 rad of alpha particles is absorbed by the body, 10 to 20 times as much biological effect is produced as there would be in the absorption of 1 rad of X rays, gamma rays, or beta particles.

Sometimes, then, one speaks of the *relative biological effectiveness* (RBE) of radiation, or the *roentgen equivalent, man* (rem). A rad of X rays, gamma rays, or beta particles has a rem of 1, while a rad of alpha particles has a rem of 10 to 20.

If we allow for the effect of the larger particles (which are not very common under ordinary conditions) we can estimate that the gonads of the average human being receive a total dose of natural radiation of about 3 rems per 30 years. This is just about an irreducible minimum.

Man-made Radiation

Man began to add to the background radiation in the 1890s. In 1895, X rays were discovered and since then have become increasingly useful in medical diagnosis and therapy and in industry. In 1896, radioactivity was discovered and radioactive substances were concentrated in laboratories in order that they might be studied. In 1934, it was found that radioactive forms of nonradioactive elements *(radioisotopes)* could be formed and their use came to be widespread in universities, hospitals, and industries.*

Then, in 1945, the nuclear bomb was developed. With the uranium or plutonium fission that produces a nuclear explosion, there is an accompaniment of intense gamma radiation. In addition, a variety of radioisotopes are left behind in the form of the residue *(fission fragments)* of the fissioning atoms. These fission fragments are distributed widely in the atmosphere. Some rise high into the stratosphere and descend (as *fallout*) over the succeeding months and years.†

It is hard to try to estimate how much additional radiation is being absorbed by human beings out of these man-made sources. Fallout is not uniformly spread over the earth but is higher in those latitudes where nuclear bombs have been most frequently tested. Then, too, people in industries and research who are involved with the use of radioisotopes, and people in medical centers who constantly deal with X rays, are likely to get more exposure than others.

These adjuncts of modern science and medicine are more common and widespread in technologically advanced countries than elsewhere, and nuclear bombs have most often been exploded in just those latitudes where the advanced countries are to be found.

Attempts have been made to work out estimates of this exposure. One estimate, involving a number of technologically advanced countries (including the United States)

---

*For more about this subject, see *Radioisotopes in Industry* and *Radioisotopes in Medicine,* companion booklets in this series.

†For more about this subject, see *Fallout from Nuclear Tests,* another booklet in this series.

showed that an average of somewhere between 0.02 and 0.18 rem per year was absorbed, as a result of radiations (usually X rays) used in medical diagnosis and therapy. Occupational exposure added, on the average, not more than 0.003 rem, though the individuals constantly exposed in the course of their work would naturally absorb considerably more than this overall average.

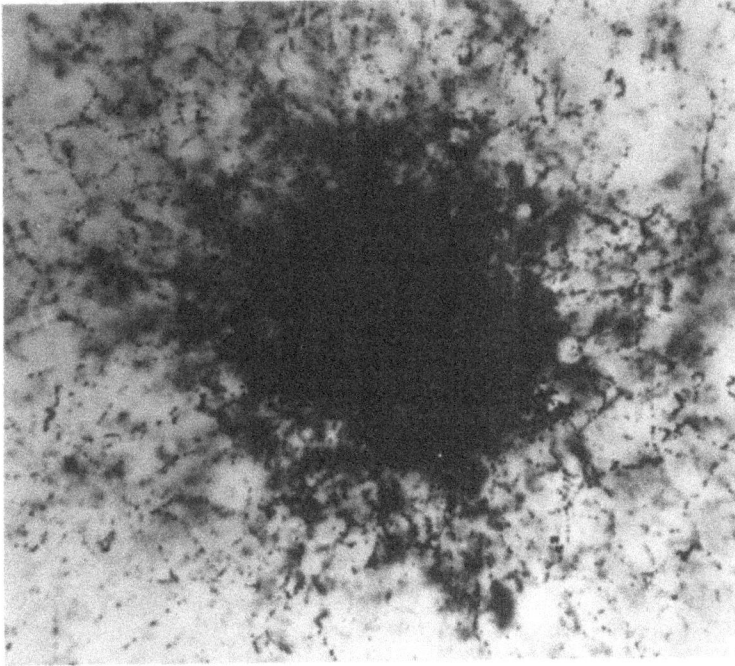

Man-made radioactivity in the atmosphere produced this nuclear-emulsion photograph. This radiation source is a fission product produced in a nuclear explosion. The enlargement is 1200 diameters. Compare this with the natural radioactivity depicted on page 28.

On the whole, the highest absorption was found, as was to be expected, in the United States.

If these findings are expanded to cover a 30-year period, assuming the absorption will remain the same from year to year, it turns out that the average absorption of man-made radiation in the nations studied varies from 0.6 rem to 5.5 rems per 30 years per individual.

Considering the higher figure to be applicable to the United States, it would seem that man-made radiation from all sources is now being absorbed at nearly twice the rate that natural radiation is. To put it another way, Americans are just about tripling their radiation dosage by reason of the human activities that are now adding man-made radiation to the natural supply. By far the major part of this additional dosage is the result of the use of X rays in searching for decayed teeth, broken bones, lung lesions, swallowed objects, and so on.

DOSE AND CONSEQUENCE

Radiation Sickness

The danger to the individual as a result of overexposure to high-energy radiation was understood fairly soon but not before some tragic experiences were recorded.

One of the early workers with radioactive materials, Pierre Curie, deliberately exposed a patch of his skin to the action of radioactive radiations and obtained a serious and slow-healing burn. His wife, Marie Curie, and their daughter, Irène Joliot-Curie, who spent their lives working with radioactive materials, both died of leukemia, very possibly as the result of cumulative exposure to radiation. Other research workers in the field died of cancer before the full necessity of extreme caution was understood.

The damage done to human beings by radiation could first be studied on a large scale among the survivors of the nuclear bombings of Hiroshima and Nagasaki in 1945. Here marked symptoms of *radiation sickness* were observed. This sickness often leads to death, though a slow recovery is sometimes possible.

In general, high-energy radiation damages the complex molecules within a cell, interfering with its chemical machinery to the point, in extreme cases, of killing it. (Thus, cancers, which cannot safely be reached with the surgeon's knife, are sometimes exposed to high-energy radiation in the hope that the cancer cells will be effectively killed in that manner.)

The delicate structure of the genes and chromosomes is particularly vulnerable to the impact of high-energy radiation. Chromosomes can be broken by such radiation and this is the main cause of actual cell death. A cell that is not killed outright by radiation may nevertheless be so damaged as to be unable to undergo replication and mitosis.

If a cell is of a type that will not, in the course of nature, undergo division, the destruction of the mitosis machinery is not in itself fatal to the organism. A creature like *Drosophila*, which, in its adult stage, has very few cell divisions going on among the ordinary cells of its body, can survive radiation doses a hundred times as great as would suffice to kill a man.

In a human being, however — even in an adult who is no longer experiencing overall growth — there are many tissues whose cells must undergo division throughout life. Hair and fingernails grow constantly, as a result of cell division at their roots. The outer layers of skin are steadily lost through abrasion and are replaced through constant cell division in the deeper layers. The same is true of the lining of the mouth, throat, stomach, and intestines. Too, blood cells are continually breaking up and must be replaced in vast numbers.

If radiation kills the mechanism of division in only some of these cells, it is possible that those that remain reasonably intact can divide and eventually replace or do the work of those that can no longer divide. In that case, the symptoms of radiation sickness are relatively mild in the first place and eventually disappear.

Past a certain critical point, when too many cells are made incapable of division, this is no longer possible. The symptoms, which show up in the growing tissues particularly (as in the loss of hair, the misshaping or loss of fingernails, the reddening and hemorrhaging of skin, the ulceration of the mouth, and the lowering of the blood cell count), grow steadily more severe and death follows.

Radiation and Mutation

Where radiation is insufficient to render a cell incapable of division, it may still induce mutations, and it is in this

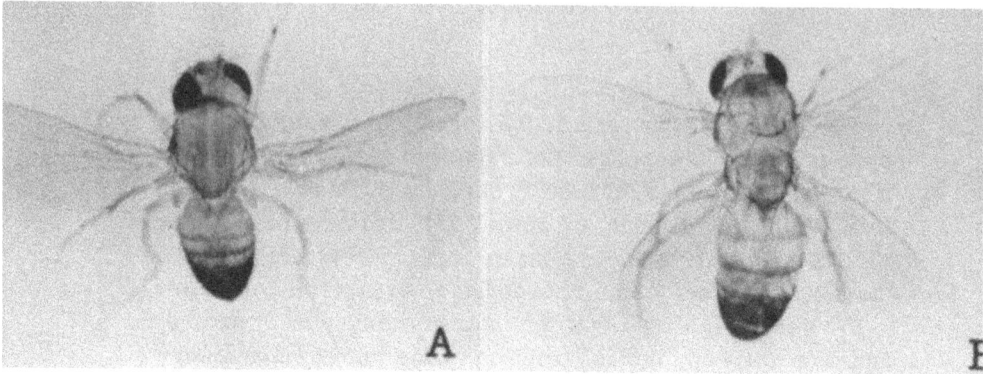

A   B

Studies at the California Institute of Technology furnish information on the nature of radiation effects on genes. The experiments produced fruit flies with three or four wings and double or partially doubled thoraxes by caus-

fashion that skin cancer, leukemia, and other disorders may be brought about.*

Mutations can be brought about in the sex cells, too, of course, and when this happens it is succeeding generations that are affected and not merely the exposed individual. Indeed, where the sex cells are concerned, the relatively mild effect of mutation is more serious than the drastic one of nondivision. A fertilized ovum that cannot divide eventually dies and does no harm; one that can divide but is altered, may give rise to an individual with one of the usual kinds of major or minor physical defects.

The effect of high-energy radiation on the genetic mechanism was first demonstrated experimentally in 1927 by Muller. Using *Drosophila* he showed that after large doses of X rays, flies experienced many more lethal mutations per chromosome than did similar flies not exposed to radiation. The drastic differences he observed proved the connection between radiation and mutation at once.

Later experiments, by Muller and by others, showed that the number of mutations was directly proportional to the quantity of radiation absorbed. Doubling the quantity of radiation absorbed doubled the number of mutations, tripling the one tripled the other, and so on. This means that if the number of mutations is plotted against the amount of radiation absorbed, a straight line can be drawn.

---

*For details on *somatic* effects of radiation, see *Your Body and Radiation*, a companion booklet in this series.

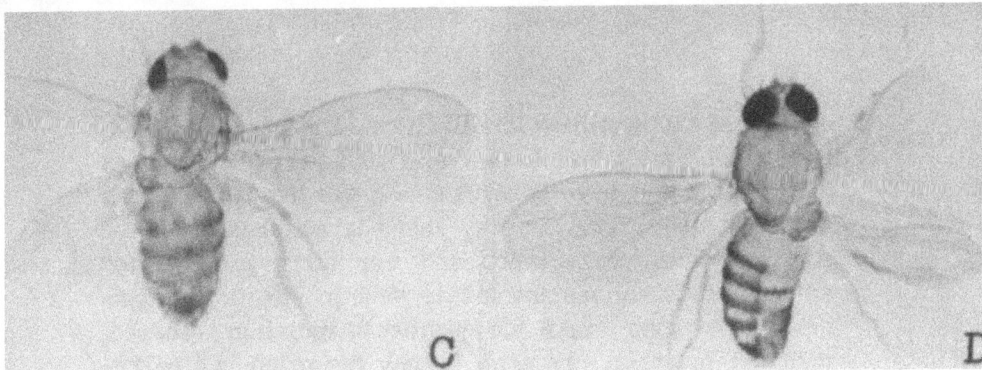

ing gene mutation through X-irradiation and chromosome rearrangements. A is a normal male Drosophila; B is a four-winged male with a double thorax; and C and D are three-winged flies with partial double thoraxes.

It is generally believed that the straight line continues all the way down without deviation to very low radiation absorptions. This means there is no "threshold" for the mutational effect of radiation.* No matter how small a dosage of radiation the gonads receive, this will be reflected in a proportionately increased likelihood of mutated sex cells with effects that will show up in succeeding generations.

In this respect, the genetic effect of radiation is quite different from the somatic effect. A small dose of radiation may affect growing tissues and prevent a small proportion of the cells of those tissues from dividing. The remaining, unaffected cells take up the slack, however, and if the proportion of affected cells is small enough, symptoms are not visible and never become visible. There is thus a threshold effect: The radiation absorbed must be more than a certain amount before any somatic symptoms are manifest.

Matters are quite different where the genetic effect is concerned. If a sex cell is damaged and if that sex cell is one of the pair that goes into the production of a fertilized ovum, a damaged organism results. There is no margin for correction. There is no unaffected cell that can take over the work of the damaged sex cell once fertilization has taken place.

Suppose only one sex cell out of a million is damaged. If so, a damaged sex cell will, on the average, take part in

---

*An apparent threshold has been found in mice, see pages 40−42.

one out of every million fertilizations. And when it is used, it will not matter that there are 999,999 perfectly good sex cells that might have been used—it was the damaged cell that *was* used. That is why there is no threshold in the genetic effect of radiation and why there is no "safe" amount of radiations insofar as genetic effects are concerned. However small the quantity of radiation absorbed, mankind must be prepared to pay the price in a corresponding increase of the genetic load.

If the straight line obtained by plotting mutation rate against radiation dose is followed down to a radiation dose of zero, it is found that the line strikes the vertical axis slightly above the origin. The mutation rate is more than zero even when the radiation dose is zero. The reason for this is that it is the dose of man-made radiation that is being considered. Even when man-made radiation is completely absent there still remains the natural background radiation.

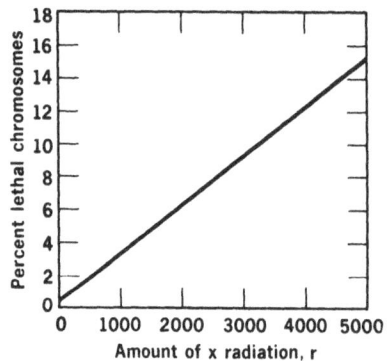

It is possible in this manner to determine that background radiation accounts for considerably less than 1% of the spontaneous mutations that take place. The other mutations must arise out of chemical misadventures, out of the random heat-jiggling of molecules, and so on. These, it can be presumed, will remain constant when the radiation dose is increased.

This is a hopeful aspect of the situation for it means that, if the background radiation is doubled or tripled for mankind as a whole, only that small portion of the spontaneous mutation rate that is due to the background radiation will be doubled or tripled.

Let us suppose, for instance, that fully 1% of the spontaneous mutations occurring in mankind is due to background radiation. In that case, the tripling of the background radiation produced in the United States by man-made causcc (see Table) would triple that 1%. In place of 99 non-radiational mutations plus 1 radiational, we would have 99 plus 3. The total number of mutations would increase from 100 to 102 —an increase of 2%, not an increase of 200% that one would expect if all spontaneous mutations were caused by background radiation.

RADIATION EXPOSURES IN THE UNITED STATES*

	Millirems†
Natural Sources	
A. External to the body	
1. From cosmic radiation	50.0
2. From the earth	47.0
3. From building materials	3.0
B. Inside the body	
1. Inhalation of air	5.0
2. Elements found naturally in human tissues	21.0
Total, Natural sources	126.0
Man-made Sources	
A. Medical Procedures	
1. Diagnostic X rays	50.0
2. Radiotherapy X ray, radioisotopes	10.0
3. Internal diagnosis, therapy	1.0
Subtotal	61.0
B. Atomic energy industry, laboratories	0.2
C. Luminous watch dials, television tubes, radioactive industrial wastes, etc.	2.0
D. Radioactive fallout	4.0
Subtotal	6.2
Total, man-made sources	67.2
Overall total	193.2

*Estimated average exposures to the gonads, based on 1963 report of Federal Radiation Council.
†One thousandth of a rem.

Dosage Rates

Another difference between the genetic and somatic effects of radiation rests in the response to changes in the rate at which radiation is absorbed. It makes a considerable difference to the body whether a large dose of radiation is absorbed over the space of a few minutes or a few years.

When a large dose is absorbed over a short interval of time, so many of the growing tissues lose the capacity for cell division that death may follow. If the same dose is delivered over years, only a small bit of radiation is absorbed on any given day and only small proportions of growing cells lose the capacity for division at any one time. The unaffected cells will continually make up for this and will replace the affected ones. The body is, so to speak, continually repairing the radiation damage and no serious symptoms will develop.

Then, too, if a moderate dose is delivered, the body may show visible symptoms of radiation sickness but can recover. It will then be capable of withstanding another moderate dose, and so on.

The situation is quite different with respect to the genetic effects, at least as far as experiments with *Drosophila* and bacteria seem to show. Even the smallest doses will produce a few mutations in the chromosomes of those cells in the gonads that eventually develop into sex cells. The affected gonad cells will continue to produce sex cells with those mutations for the rest of the life of the organism. Every tiny bit of radiation adds to the number of mutated sex cells being constantly produced. There is no recovery, because the sex cells, after formation, do not work in cooperation, and affected cells are not replaced by those that are unaffected.

This means (judging by the experiments on lower creatures) that what counts, where genetic damage is in question, is not the rate at which radiation is absorbed but the total sum of radiation. Every exposure an organism experiences, however small, adds its bit of damage.

Accepting this hard view, it would seem important to make every effort to minimize radiation exposure for the population generally.

Since most of the man-made increase in background radiation is the result of the use of X rays in medical diagnosis and therapy, many geneticists are looking at this with suspicion and concern. No one suggests that their use be abandoned, for certainly such techniques are important in the saving of life and the mitigation of suffering. Still, X rays ought not to be used lightly, or routinely as a matter of course.

It might seem that X rays applied to the jaw or the chest would not affect the gonads, and this might be so if all the X rays could indeed be confined to the portion of the body at which they are aimed. Unfortunately, X rays do not uniformly travel a straight line in passing through matter. They are scattered to a certain extent; if a stream of X rays passes through the body anywhere, or even through objects near the body, some X rays will be scattered through the gonads.

It is for this reason that some geneticists suggest that the history of exposure to X rays be kept carefully for each person. A decision on a new exposure would then be determined not only by the current situation but by the individual's past history.

Such considerations were also an important part of the driving force behind the movement to end atmospheric testing of nuclear bombs. While the total addition to the background radiation resulting from such tests is small, the prospect of continued accumulation is unpleasant.

What's more, whereas X rays used in diagnosis and therapy have a humane purpose and chiefly affect the patient who hopes to be helped in the process, nuclear fallout affects all of humanity without distinction and seems, to many people, to have as its end only the promise of a totally destructive nuclear war.

It is not to be expected that the large majority of humanity that makes up the populations outside the United States, Great Britain, France, China, and the Soviet Union can be expected to accept stoically the risk of even limited quantities of genetic damage, out of any feeling of loyalty to nations not their own. Even within the populations of the three major nuclear powers there are strong feelings that the possible benefits of nuclear testing do not balance the certain dangers.

Public opinion throughout the world is a key factor, then, in enforcing the Nuclear Test Ban Treaty, signed by the governments of the United States, Great Britain, and the Soviet Union on October 10, 1963.

Effects on Mammals

Although genetic findings on such comparatively simple creatures as fruit flies and bacteria seem to apply generally to all forms of life, it seems unsafe to rely on these findings completely in anything as important as possible genetic damage to man through radiation. During the 1950s and 1960s, therefore, there have been important studies on mice, particularly by W. L. Russell at Oak Ridge National Laboratory, Oak Ridge, Tennessee.

While not as short-lived or as fecund as fruit-flies, mice can nevertheless produce enough young over a reasonable period of time to yield statistically useful results. Experimenters have worked with hundreds of thousands of offspring born of mice that have been irradiated with gamma rays and X rays in different amounts and at different intensities, as well as with additional hundreds of thousands born to mice that were not irradiated.

Since mice, like men, are mammals, results gained by such experiments are particularly significant. Mice are far closer to man in the scheme of life than is any other creature that has been studied genetically on a large scale, and their reactions (one might cautiously assume) are likely to be closer to those that would be found in man.

Almost at once, when the studies began, it turned out that mice were more susceptible to genetic damage than fruit flies were. The induced mutation rate per gene seems to be about fifteen times that found in *Drosophila* for comparable X ray doses. The only safe course for mankind then is to err, if it must, strongly on the side of conservatism. Once we have decided what might be safe on the basis of *Drosophila* studies, we ought then to tighten precautions several notches by remembering that we are very likely more vulnerable than fruit flies are.

Counteracting the depressing nature of this finding was that of a later, quite unexpected discovery. It was well established that in fruit flies and other simple organisms, it was the total dosage of absorbed radiation that counted and that whether this was delivered quickly or slowly did not matter.

Arrangement for long-term low-dose-rate irradiation of mice used for mutation-rate studies at Oak Ridge National Laboratory. The cages are arranged at equal distances from a cesium-137 gamma-ray source in the lead pot on the floor. The horizontal rod rotates the source.

This proved to be *not* so in the case of mice. In male mice, a radiation dose delivered at the rate of 0.009 rad per minute produced only from one-quarter to one-third as many mutations as did the same total dose delivered at 90 rads per minute.

In the male, cells in the gonads are constantly dividing to produce sex cells. The latter are produced by the billions. It might be, then, that at low radiation dose rates, a few of the gonad cells are damaged but that the undamaged ones produce a flood of sperm cells, "drowning out" the few produced by the damaged gonad cells. The same radiation dose delivered in a short time might, however, damage so many of the gonad cells as to make the damaged sex cells much more difficult to "flood out".

A second possible explanation is that there is present within the cells themselves some process that tends to repair damage to the genes and to counteract mutations. It might be a slow-working, laborious process that could keep up with the damage inflicted at low dosage rates but not at high ones. High dosage rates might even damage the repair mechanism itself. That, too, would account for the fewer mutations at low dosage rates than at high ones.

To check which of the two possible explanations was nearer the truth, Russell performed similar tests on female mice. In the female mouse (or the female human being, for that matter) the egg cells have completed almost all their divisions before the female is born. There are only so many cells in the female gonads that can give rise to egg cells, and each one gives rise to only a single egg cell. There is no possibility of damaged egg cells being drowned out by floods of undamaged ones because there are no floods.

Yet it was found that in the female mouse the mutation rate also dropped when the radiation dose rate was decreased. In fact, it dropped even more drastically than was the case in the male mouse.

Apparently, then, there must be actual repair within the cell. There must be some chemical mechanism inside the cell capable of counteracting radiation damage to some extent. In the female mouse, the mutation rate drops very low as the radiation dose rate drops, so that it would seem that almost all mutations might be repaired, given enough time. In the male, the mutation rate drops only so far and no farther, so that some mutations (about one-third is the best estimate so far) cannot be repaired.

If this is also true in the human being (and it is at least reasonably likely that it is), then the greater vulnerability of our genes as compared with those of fruit flies is at least partially made up for by our greater ability to repair the damage.

This opens a door for the future, too. The workings of the gene-repair mechanism ought (it is to be hoped) eventually to be puzzled out. When it is, methods may be discovered for reinforcing that mechanism, speeding it, and increasing its effectiveness. We may then find ourselves no longer completely helpless in the face of genetic damage, or even of radiation sickness.

On the other hand, it is only fair to point out that the foregoing appraisal may be an over-optimistic view. Russell's experiments involved just 7 genes and it is possible that these are not representative of the thousands that exist altogether. While the work done so far is most suggestive and interesting, much research remains to be carried out.

If, then, we cannot help hoping that natural devices for counteracting radiation damage may be developed in the future, we must, for the present, remain rigidly cautious.

Conclusion

It is unrealistic to suppose that all sources of man-made radiation should be abolished. The good they do now, the greater good they will do in the future, cannot be abandoned. It is, however, reasonable to expect that the present Nuclear Test Ban Treaty will continue and that nations, such as France and China, which have nuclear capabilities but are not signatories of the Treaty will eventually sign. It is also reasonable to expect that X ray diagnosis and therapy will be carried on with the greatest circumspection, and that the use of radiation in industry and research will be carried on with great care and with the use of ample shielding.

A film badge (left) and a personal radiation monitor (right) record the amount of radiation absorbed by the wearer. These safety devices, worn by persons working in radiation environments, are designed to keep a constant check on each individual's absorbed dose and to prevent overexposure.

As long as man-made radiation exists, there will be some absorption of it by human beings. The advantages of its use in our modern society are such that we must be prepared to pay some price. This is not a matter of callousness. We have come to depend a great deal for comfort and even for extended life, upon the achievements of our technology, and any serious crippling of that technology will cost us lives. An attempt must be made to balance the values of radiation against its dangers; we must balance lives against lives. This involves hard judgments.

Those working under conditions of greatest radiation risk—in atomic research, in industrial plants using isotopes, and so on—can be allowed to set relatively high limits for total radiation dosages and dose rates that they may absorb (with time) with reasonable safety, but such rates will never do for the population generally. A relative few can voluntarily endure risks, both somatic and genetic, that we cannot sanely expect of mankind as a whole.*

From fruit fly experiments it would seem that a total exposure of 30 to 100 rads of radiation will double the spontaneous mutation rate. So much radiation and such a doubling of the rate would be considered intolerable for humanity.

Some geneticists have recommended that the average total exposure of human beings in the first 30 years of life be set at 10 rads. Note that this figure is set as a *maximum*. Every reasonable method, it is expected, will be used to allow mankind to fall as far short of this figure as possible. Note also that the 10-rad figure is an *average* maximum. The exposure of some individuals to a greater total dose would be viewed as tolerable for society if it were balanced by the exposure of other individuals to a lesser total dose.

A total exposure of 10 rads might increase the overall mutation rate, it is roughly estimated, by 10%. This is serious enough, but is bearable if we can convince our-

---

*Nevertheless, it should be pointed out that the precautions taken in the atomic energy industry are such that absorption of radiation is not as severe a problem as one might suspect. Fully 95% of those engaged in this work receive less than 1 rem a year. Only 1% receive more than 5 rems.

selves that the alternative of abandoning radiation technology altogether will cause still greater suffering.

A 10% increase in mutation rate, whatever it might mean in personal suffering and public expense, is not likely to threaten the human race with extinction, or even with serious degeneration.

The human race as a whole may be thought of as somewhat analogous to a population of dividing cells in a growing tissue. Those affected by genetic damage drop out and the slack is taken up by those not affected.

If the number of those affected is increased, there would come a crucial point, or threshold, where the slack could no longer be taken up. The genetic load might increase to the point where the species as a whole would degenerate and fade toward extinction—a sort of "racial radiation sickness".

We are not near this threshold now, however, and can, therefore, as a species, absorb a moderate increase in mutation rate without danger of extinction.

On the other hand, it is *not* correct to argue, as some do, that an increase in mutation rate might be actually beneficial. The argument runs that a higher mutation rate might broaden the gene pool and make it more flexible, thus speeding up the course of evolution and hastening the advent of "supermen"—brainier, stronger, healthier than we ourselves are.

The truth seems to be that the gene pool, as it exists now, supplies us with all the variability we need for the effective working of the evolutionary mechanism. That mechanism is functioning with such efficiency that broadening the gene pool cannot very well add to it, and if the hope of increased evolutionary efficiency were the only reason to tolerate man-made radiation, it would be insufficient.

The situation is rather analogous to that of a man who owns a good house that is heavily mortgaged. If he were offered a second house with a similar mortgage, he would have to refuse. To be sure, he would have twice the number of houses, but he would not need a second house since he has all the comfort he can reasonably use in his first house—and he would not be able to afford a second mortgage.

What humanity must do, if additional radiation damage is absolutely necessary, is to take on as little of that added damage as possible, and not pretend that any direct benefits will be involved. Any pretense of that sort may well lure us into assuming still greater damage—damage we may not be able to afford under any circumstances and for any reason.

Actually, as the situation appears right now, it is not likely that the use of radiation in modern medicine, research, and industry will overstep the maximum bounds set by scientists who have weighed the problem carefully. Only nuclear warfare is likely to do so, and apparently those governments with large capacities in this direction are thoroughly aware of the danger and (so far, at least) have guided their foreign policies accordingly.

SUGGESTED REFERENCES

Books

Radiation, Genes, and Man, Bruce Wallace and Theodosius Dobzhansky, Holt, Rinehart and Winston, Inc., New York 10017, 1963, 205 pp., $5.00(hardback); $1.28 (paperback).

Genetics in the Atomic Age (second edition), Charlotte Auerbach, Oxford University Press, Inc., Fair Lawn, New Jersey 07410, 1965, 111 pp., $2.50.

Atomic Radiation and Life (revised edition), Peter Alexander, Penguin Books, Inc., Baltimore, Maryland 21211, 1966, 288 pp., $1.65.

The Genetic Code, Isaac Asimov, Grossman Publishers, Inc., The Orion Press, New York 10003, 1963, 187 pp., $3.95 (hardback); $0.60 (paperback) from the New American Library of World Literature, Inc., New York 10022.

Radiation: What It Is and How It Affects You. Ralph E. Lapp and Jack Schubert, The Viking Press, New York 10022, 1957, 314 pp., $4.50 (hardback); $1.45 (paperback).

Report of the United Nations Scientific Committee on the Effects of Atomic Radiation, General Assembly, 19th Session, Supplement No. 14 (A/5814), United Nations, International Documents Service, Columbia University Press, New York 10027, 1964, 120 pp., $1.50.

The Effects of Nuclear Weapons, Samuel Glasstone (Ed.), U. S. Atomic Energy Commission, 1962, 730 pp., $3.00. Available from the Superintendent of Documents, U. S. Government Printing Office, Washington, D. C. 20402.

Effect of Radiation on Human Heredity, World Health Organization, International Documents Service, Columbia University Press, New York 10027, 1957, 168 pp., $4.00.

The Nature of Radioactive Fallout and Its Effects on Man, Hearings before the Special Subcommittee on Radiation of the Joint Committee on Atomic Energy, Congress of the United States, 85th Congress, 1st Session, U. S. Government Printing Office, 1957, Volume I, 1008 pp., $3.75; Volume II, 1057 pp., $3.50. Available from the Office of the Joint Committee on Atomic Energy, Congress of the United States, Senate Post Office, Washington, D. C. 20510.

Genetics, Radiobiology, and Radiology, Proceedings of the Midwestern Conference, Wendell G. Scott and Evans Titus, Charles C Thomas Publisher, Springfield, Illinois 62703, 1959, 166 pp., $5.50.

Articles

Genetic Hazards of Nuclear Radiations, Bentley Glass, *Science,* 126: 241 (August 9, 1957).

Genetic Loads in Natural Populations, Theodosius Dobzhansky, *Science,* 126: 191 (August 2, 1957).

Radiation Dose Rate and Mutation Frequency, W. L. Russell and others, *Science,* 128: 1546 (December 19, 1958).

Ionizing Radiation and the Living Cell, Alexander Hollaender and George E. Stapleton, *Scientific American,* 201: 95 (September 1959).

Radiation and Human Mutation, H. J. Muller, *Scientific American,* 193: 58 (November 1955).

Ionizing Radiation and Evolution, James F. Crow, *Scientific American,* 201: 138 (September 1959).

Motion Pictures

Radiation and the Population, 29 minutes, sound, black and white, 1962. Produced by the Argonne National Laboratory. This film explains how radiation causes mutations and how these mutations are passed on to succeeding generations. Mutation research is illustrated with results of experimentation on generations of mice. A discussion of work with fruit flies and induced mutations is also included. This film is available for loan without charge from the AEC Headquarters Film Library, Division of Public Information, U. S. Atomic Energy Commission, Washington, D. C. 20545 and from other AEC film libraries.

Mutation, 28 minutes, sound, color, 1962. Produced by the American Institute of Biological Sciences and may be rented from the Text-Film Division, McGraw-Hill Book Company, 330 West 42nd Street, New York 10036. This film discusses chromosomal and genetic mutations as applied to man. Muller's work in inducing mutations by X rays is described.

These three films are 30 minutes long, have sound, are in black and white, and were released in 1960. They are part of a 48-film series that is correlated with the textbook, *Principles of Genetics,* (fifth edition), Edmund W. Sinnott, L. C. Dunn, and Theodosius Dobzhansky, McGraw-Hill Book Company, 1958, 459 pp., $8.50.

Mutagen-Induced Gene Mutation. The narrator of this film is Hermann J. Muller, who won a Nobel Prize in 1946 for his work in the field of genetics. The measurement of X-ray dose in roentgens and the dose required to double the spontaneous mutation rate in *Drosophila* and mice are discussed. The magnitude and meaning of permissible doses of high-energy radiation are discussed. Other mutagenic agents (ultraviolet light and chemical substances) are discussed, concluding with comments on the importance of gene mutation in the present and future.

Selection, Genetic Death and Genetic Radiation Damage. The narrator of this film is Theodosius Dobzhansky, the coauthor of this booklet. Genetic death is discussed in detail, as are examples of how genetic loads are changed subsequent to radiation exposure. While it is generally agreed that the great majority of mutants are harmful when homozygous, more evidence is needed about the beneficial and detrimental effects of mutants when heterozygous. In the case of sickle cell anemia, heterozygotes are adaptively superior to normal homozygotes. This makes for balanced polymorphism, by which a gene is retained in the population despite its lethality when homozygous because of the advantage it confers when heterozygous.

Gene Structure and Gene Action. The lecturer of this film is G. W. Beadle of Cornell University. The Watson-Crick structure of DNA is discussed in terms of mutation. Several tests of the chain separation hypothesis for DNA replication are described (experiments with heavy DNA, radioactive chromosomes, and the replication of DNA in vitro). This working hypothesis is presented: The coded information in DNA is transferred to RNA, which serves as a template for polypeptide synthesis.

PHOTO CREDITS

Dr. Asimov's photograph by David R. Phillips, courtesy *Chemical and Engineering News*

Page

4   James German, M.D.
6   Bausch & Lomb, Inc.
12   James German, M.D.
20   Indiana University
24   Robert C. Filz, Air Force Cambridge Research Laboratories
25   J. K. Bøggild, Niels Bohr Institute, Copenhagen University
26   Brookhaven National Laboratory
28,31 Herman Yagoda, Air Force Cambridge Research Laboratories
41   Oak Ridge National Laboratory

www.ingramcontent.com/pod-product-compliance
Lightning Source LLC
Chambersburg PA
CBHW032020190326
41520CB00007B/551